U.S.NRC
United States Nuclear Regulatory Commission
Protecting People and the Environment

Design Control

In Pursuit of Engineering Excellence

A Quick Reference Guide
for NRC Inspectors

Region III

"All men by nature desire knowledge."

Aristotle

"The important thing is not to stop questioning. Curiosity has its own reason for existing."

Albert Einstein

FOREWORD

Knowledge management and transfer have become increasingly important as the U.S. Nuclear Regulatory Commission seeks to share the vast inspection knowledge of its experienced inspectors with those who have recently joined the agency. This booklet should serve as an inspection reference to further increase the understanding by, and development of, NRC inspectors who perform inspections in the areas covered by this booklet.

The contents of this design control quick reference booklet resulted from a collaboration of all four regional offices in support of the agency's focus on knowledge transfer, with staff from various other NRC offices providing valuable comments and suggestions. The principal and contributing authors of this booklet are listed below.

Julio Lara, P.E., Region III
Jennifer Tifft, Region I
Frank Arner, Region I
Randy Moore, Region II
George Replogle, Region IV

Note that the guidance contained in this booklet is not intended to be all inclusive but rather to supplement existing inspection procedures, to heighten inspectors' awareness, and further enhance the effectiveness of engineering inspections. The various inspection manual chapters and other regulatory documents discussed herein, provide the official agency inspection policies and guidance.

Appendix B, "Quality Assurance Criteria for Nuclear Power Plants and Fuel Reprocessing Plants," to Title 10, Part 50, "Domestic Licensing of Production and Utilization Facilities," of the Code of *Federal Regulations* (10 CFR Part 50)

Criterion III, Design Control

Measures shall be established to assure that applicable regulatory requirements and the design basis, as defined in 10 CFR 50.2 and as specified in the licensee application, for those structures, systems, and components to which this Appendix applies are correctly translated into specifications, drawings, procedures, and instructions. These measures shall include revisions to assure that appropriate quality standards are specified and included in design documents and that deviations from such standards are controlled. Measures shall also be established for the selection and review for suitability of application of materials, parts, equipment, and processes that are essential to the safety-related functions of the structures, systems, and components.

Measures shall be established for the identification and control of design interfaces and for coordination among participating design organizations. These measures shall include the establishment of procedures among participating design organizations for the review, approval, release, distribution, and revision of documents involving design interfaces.

The design control measures shall provide for verifying or checking the adequacy of design, such as by the performance of design reviews, by the use of alternate or simplified calculational methods, or by the performance of a suitable testing program. The verifying or checking process shall be performed by individuals or groups other than those who performed the original design, but who may be from the same organization. Where a test program is used to verify the adequacy of a specific design feature in lieu of other verifying or checking processes, it shall include suitable qualification testing of a prototype unit under the most adverse design conditions. Design control measures shall be applied to items such as the following: reactor physics, stress, thermal hydraulic, and accident analysis; compatibility of materials; accessibility for inservice inspection, maintenance and repair; the delineation of acceptance criteria for inspections and tests.

Design changes, including field changes, shall be subject to design control measures commensurate with those applied to the original design and be approved by the organization that performed the original design unless the applicant designates another responsible organization.

CONTENTS

FOREWORD i

PURPOSE 1

DEFINITIONS 2

GUIDANCE FOR ENGINEERING INSPECTIONS 5

    A.   NRC Inspections Involving Engineering Design Activities 5

    B.   10 CFR 50.55a, "Codes and Standards" 5

    C.   Standard Review Plans 7

    D.   Systematic Evaluation Program 8

    E.   Safety Evaluation Reports 9

    F.   Technical Guidance—Manual Chapter Part 9900 9

DESIGN AND LICENSING BASIS 11

    A.   Design Basis (Bases) 11

    B.   Licensing Basis 12

DESIGN CONTROL GUIDANCE DOCUMENTS 15

CRITERION III OF APPENDIX B TO 10 CFR PART 50 CROSS-REFERENCE TABLE 19

PLANT ENGINEERING PRODUCTS 21

    A.   Overview 21

    B.   Calculations 21

    C.   Modifications and Temporary Modifications 24

    D.   Operability Determinations 25

    E.   Procedures 26

    F.   Corrective Action Determinations 27

    G.   Dedication of Commercial-Grade Parts for Safety-Related Applications 27

ROBUST APPLICATION OF ENGINEERING PRINCIPLES 29

VALUE-ADDED FINDINGS AND STARS DOCUMENTS 31

REFERENCES 33

ACRONYMS 37

PURPOSE

This booklet was developed primarily for new U.S. Nuclear Regulatory Commission (NRC) inspectors who conduct engineering-focused inspections. This booklet may also serve as a quick reference for other experienced inspectors. In user-friendly language, it provides inspection guidance and contains useful inspection tips.

The main purpose of the booklet is to develop a fundamental understanding of what constitutes design control, as required by Criterion III, "Design Control," of Appendix B, "Quality Assurance Criteria for Nuclear Power Plants and Fuel Reprocessing Plants," to Title 10, Part 50, "Domestic Licensing of Production and Utilization Facilities," of the Code of Federal Regulations (10 CFR Part 50). The booklet structure begins with basic terminology used in the nuclear industry to ensure a common understanding. NRC engineering inspectors frequently rely on industry codes and standards and on NRC guidance documents. The booklet discusses the importance of these documents and how inspectors can use them. Because inspectors must develop a sound understanding of design and licensing basis documents, these documents are introduced. After touching on engineering terminology, codes and standards, and design and licensing basis, the booklet then delves into Criterion III of Appendix B to 10 CFR Part 50 to help inspectors understand the various requirements of the criterion and into NRC and industry guidance documents regarding the various elements of design control. Licensee engineering design programs and activities should ensure compliance with the requirements in Criterion III with respect to engineering work products, such as calculations, plant modifications, engineering analysis, and procurement and dedication. The booklet discusses these various engineering work products to enhance inspectors' knowledge and understanding. Lastly, the booklet briefly explores the meaning of the commonly used term "robust application of engineering principles."

DEFINITIONS

The definitions below came from various sources, including NRC regulations and guidance documents and industry standards, and reflect generally accepted engineering terminology. However, because a number of these terms are not explicitly defined in NRC regulations, some variations may exist. Documents used to develop this guidance booklet are listed in the References section.

Assumptions
Assumptions are statements that define premises, limitations, or restrictions that are accepted as true without a thorough demonstration.

Commercial-Grade Dedication
Commercial-grade dedication is an acceptance process that provides reasonable assurance that a commercial-grade item designated for use as a basic component will perform its intended safety function and, in this respect, is deemed equivalent to an item designed and manufactured under a 10 CFR Part 50, Appendix B, Quality Assurance Program. This assurance is achieved when a third-party dedicating entity or the purchaser of a commercial-grade item identifies the critical characteristics of the item and verifies the acceptability of these characteristics through inspections, tests, or analyses.

Configuration Management
Configuration management is an integrated management process used to ensure that the licensee maintains the plant's physical and functional characteristics in conformance with its design and licensing basis; that operating, training, modification, and maintenance processes are consistent with the conditions prescribed by the design and current licensing basis; and that the licensee operates and maintains the plant within these conditions.

Critical Characteristics
Critical characteristics are those important design, material, and performance characteristics of a commercial-grade item that, once verified, will provide reasonable assurance that the item will perform its intended safety function(s).

Defense in Depth
Defense in depth is a design and operational philosophy for nuclear facilities that calls for multiple layers of protection to prevent accidents and to mitigate the effects of accidents. Defense in depth includes the use of controls, multiple physical barriers to prevent release of radiation, redundant and diverse system functions, and emergency response measures.

Degraded Condition
A degraded condition is one in which the qualification of a structure, system, or component (SSC) or its functional capability is reduced. Examples of degraded conditions are failures, malfunctions, deficiencies, deviations, and defective material and equipment. Examples of conditions that can reduce the capability of a system are aging, erosion, corrosion, improper operation, and maintenance.

Design Bases

Design bases is the information that identifies the specific functions that an SSC is to perform and the specific values or ranges of values chosen for controlling parameters as reference bounds for design. These values may be (1) restraints derived from generally accepted "state-of-the-art" practices for achieving functional goals or (2) requirements derived from analysis (based on calculations and/or experiments) of the effects of a postulated accident for which an SSC must meet its functional goals.

Design Change

A design change is a change to a final design that affects the performance of an SSC.

Design Input

Design input includes those criteria, parameters, bases, or other design information upon which the final design is based. This can include technical information, design bases, performance criteria, regulatory requirements, codes, standards, analysis, and calculations.

Design Output

Design output includes documents such as drawings, specifications, and other documents that define the technical requirements of SSCs.

Design Review

Design review refers to independent verification that ensures that the licensee has incorporated important design, material, and performance characteristics into the plant design to ensure that it provides or meets safety functions while providing layers of protection to prevent and mitigate accidents.

Engineering Judgment

An engineering judgment is a determination based on prior examples, experience, or observation that has not been subjected to rigorous engineering validation.

Nonconforming Condition

A nonconforming condition is a condition of an SSC that involves a failure to meet the current licensing basis (CLB) or a situation in which the quality of an SSC is reduced because of factors such as improper design, testing, construction, or modification.

Operable

An SSC shall be operable or have operability when it is capable of performing its specified function. This includes adequate performance of any support instrumentation, controls, electrical power, cooling or seal water, lubrication or auxiliary equipment.

Part 21 (10 CFR Part 21)

In 10 CFR Part 21, "Reporting of Defects and Noncompliance," the NRC requires a responsible officer of a firm that supplies, or a facility that receives, non-complying parts and components, or defective components that cause a significant safety hazard, to immediately notify the Commission.

Qualification

Qualification refers to documented evidence that shows that a facility conforms to all aspects of design basis, including codes, standards, design criteria, and Quality Assurance (QA) regulations.

Quality Assurance
Quality Assurance comprises all those planned and systematic actions necessary to provide adequate confidence that an SSC will perform satisfactorily in service. Attributes of a QA program include programs that preserve quality through procedures, recordkeeping, inspections, corrective actions, and audits.

Redundancy
Redundancy is an alternate, independent, or duplicate method of fulfilling a safety function to mitigate the consequences of a design-basis accident.

Safety-Related Function
A safety-related function applies to the SSCs, procedures, and controls of a facility or process that must remain functional during and following design-basis events to ensure the integrity of the facility's reactor coolant pressure boundary, the facility's capability to shut down the reactor and maintain it in a safe shutdown condition, or the facility's capability to prevent or mitigate the consequences of accidents that could result in potential offsite exposure comparable to the guidelines in 10 CFR 50.34(a)(1), 10 CFR 50.67(b)(2), or 10 CFR 100.11, "Determination of Exclusion Area, Low Population Zone, and Population Center Distance." An example of a safety-related function is a facility's capability to shut down a nuclear reactor and maintain it in a safe shutdown condition.

Single Failure
A single failure is an occurrence that results in the loss of a component's capability to perform its intended safety functions. Multiple failures resulting from a single occurrence are considered to be a single failure. Fluid and electric systems are considered to be designed against an assumed single failure if neither of the following failures result in a loss of the system's capability to perform its safety functions:

(1) a single failure of any active component (assuming that passive components function properly)

(2) a single failure of a passive component (assuming that active components function properly)

Structures, Systems, and Components (safety-related)
Nuclear power plants are designed with SSCs that prevent or mitigate the consequences of postulated accidents which could cause undue risk to the health and safety of the public. See also Safety-Related Function.

Verification
Verification refers to the process of checking that the information contained in design-basis documents has been correctly and consistently translated from the source documents.

GUIDANCE FOR ENGINEERING INSPECTIONS

A.   NRC Inspections Involving Engineering Design Activities

The Reactor Oversight Process (ROP) includes several engineering-focused inspections. Engineering remains an important focal point of certain inspections because plant performance indicators do not always capture design inadequacies. Additionally, engineering can be a difficult area to inspect as design margins can often be compromised without this being readily apparent. Engineering-focused inspections allow inspectors to review equipment design and operation to verify that normal operation and routine surveillance testing assure equipment functionality under worst-case accident conditions.

Engineering-focused inspections rely on a review of a licensee's engineering work activities and products and remain an important part of the ROP. In the early to mid-1990s, the NRC's findings during inspections and reviews identified broad programmatic weaknesses that were the result of design and configuration deficiencies at some plants. These deficiencies impacted the operability of required equipment and indicated discrepancies between a plant's Updated Final Safety Analysis Report (UFSAR) and the as-built or as-modified plant or plant operating procedures. The current ROP engineering inspections allow the NRC to monitor the licensee's continued effectiveness in maintaining configuration control. To perform effective inspections, NRC inspectors need to develop and maintain a fundamental knowledge of requisite engineering quality standards and practices.

In summary, the overall goal of engineering inspections is to verify through a review of engineering work activities and products that the plant design and analyses, including calculations of record, reflect the licensing basis such as the UFSAR, Technical Specifications (TS), and other licensing documents. Furthermore, these engineering work products should be consistent with the physical plant equipment and how this equipment is operated and maintained. Consistency among the engineering and licensing basis documents, when complemented with well-operated and -maintained equipment, results in sound configuration management at a licensed facility.

B.   10 CFR 50.55a, "Codes and Standards"

Codes and standards are an integral part of the NRC's regulatory process because they provide detailed requirements and guidance to implement the NRC's broad, general design criteria prescribed in 10 CFR 50, Appendix A, "General Design Criteria for Nuclear Power Plants." The NRC and nuclear industry use codes and standards to provide greater assurance of safe plant design and operations. The NRC endorses codes and standards through Regulatory Guides (RGs) (see References Section) and in regulations. In <u>10 CFR 50.55a, "Codes and Standards,"</u> the NRC endorses Sections III and XI of the American Society of Mechanical Engineers (ASME) Boiler and Pressure Vessel (BPV) Code and the Institute of Electrical and Electronics Engineers (IEEE) Standards 279, "Criteria for Protection Systems for Nuclear Power Generating Stations," and 603, "Criteria for Safety Systems for Nuclear Power Generating Stations."

ASME Code

The ASME BPV Code contains 11 sections (I–XI), two of which are nuclear sections. The nuclear sections are Section III, which contains rules for the design and fabrication of nuclear power plant components, and Section XI, which deals with the inservice inspection (e.g., maintenance of nuclear power plant components) and inservice testing of nuclear plant components. Both of these industry standard sections constitute regulatory requirements under 10 CFR 50.55a.

The ASME Code (also referred to as the "BPV Code" or simply "the Code") is periodically revised to update the provisions for design, construction, and inservice inspection of pressure boundary components. Inspectors who perform inspections in these areas should identify which version of the Code licensees have committed to implement as part of the inspection efforts.

The BPV Code also contains "Code Cases." A Code case provides an alternative to the BPV Code and is specifically written as an alternative to a specific paragraph of the Code. The NRC's endorsement of these Code Cases is documented in RG 1.147, "Inservice Inspection Code Case Acceptability, ASME Section XI, Division 1," whereas Code Cases that the NRC has not endorsed are described in RG 1.193, "ASME Code Cases Not Approved for Use."

During access to contaminated and high radiation areas, personal safety must also be a priority.

Code interpretations are also contained within the BPV Code and differ from Code cases. These interpretations are clarifications to the BPV Code. However, these interpretations are part of neither the NRC's regulations nor its endorsed RGs, and hence the NRC is not bound by such.

Within the ASME BPV Code sections, three Code classes pertain to nuclear power plant components—Classes 1, 2, and 3. RG 1.26, "Quality Group Classifications and Standards for Water-, Steam-, and Radioactive-Waste-Containing Components of Nuclear Power Plants," describes a quality classification system related to the ASME standard that may be used to determine quality standards acceptable to the NRC staff for satisfying requirements for safety-related components in light-water-cooled nuclear power plants. Table 1 defines the three Code classes.

A separate subsection in Section XI describes the periodic inspection and repair requirements for each Code class. Many plants were built before the first issuance of Section XI. Thus, the design of the plant may prevent inspectors from examining specific component locations. Therefore, 10 CFR 50.55a allows the licensee to request relief from those Section XI Code requirements that they cannot meet. The licensee must typically justify why the examination is limited and provide an alternative examination that affords a comparable level of assurance.

Table 1. Code Classes for Nuclear Power Plant Components

Class	Discussion
1	Class 1 (Code category NB) components consist of the pressure-retaining boundary for the reactor coolant system, which generally includes all reactor coolant system connections to the outermost containment isolation valve or to the second of two normally shut valves. In 10 CFR 50.55a(c)(1), the NRC specifies that components within these boundaries must meet Class 1 requirements.
2	Class 2 (Code category NC) components consist of the pressure-retaining boundary for emergency core cooling systems (ECCS) relied on to mitigate an accident.
3	Class 3 (Code category ND) components consist of the pressure-retaining portion of cooling water systems that support the reactor shutdown function, the ECCS function, residual heat removal system functions, or spent fuel pool cooling functions.

IEEE Standards

IEEE 279 and IEEE 603 are incorporated in 10 CFR 50.55a(h). Both of these industry standards provide requirements for reactor protection systems. As with other industry standards, most IEEE standards are not explicitly endorsed through regulations but rather through RGs. The NRC issues RGs to describe methods that the staff considers acceptable for use in implementing specific parts of the agency's regulations, to explain techniques that the staff uses in evaluating specific problems or postulated accidents, and to provide guidance to applicants and licensees. RGs are not substitutes for regulations, and compliance with RGs is not required, unless explicitly incorporated into the facility's operating license (see References Section for listing of commonly applied RGs).

For nuclear power plants with construction permits issued after January 1, 1971, but before May 13, 1999, reactor protection systems must meet the requirements stated in either IEEE-279 or IEEE-603. For nuclear power plants with construction permits issued before January 1, 1971, reactor protection systems must be consistent with their licensing basis. Inspectors should review the licensee's licensing-basis documents to identify the specific commitments and requirements.

C. Standard Review Plans

In 1975, the NRC issued a Standard Review Plan (SRP) to define the scope of review and acceptance criteria for the NRC's approval of safety analysis reports. The NRC has a number of SRPs for staff use in reviewing proposed licensing actions. These actions may relate to the construction or operation of a nuclear facility or to the possession or use of nuclear materials. For nuclear power plants, the NRC has a comprehensive SRP (see NUREG-0800 at http://www.nrc.gov/reading-rm/doc-collections/nuregs/staff/sr0800/).

The Office of Nuclear Reactor Regulation (NRR) and the Office of New Reactors (NRO) are the primary users of SRPs; however, inspectors may find them useful when preparing for an inspection. SRPs can give inspectors insights into the standard that NRR uses to evaluate licensing actions.

Inspectors should keep in mind that SRPs are not regulations, and therefore they cannot propose enforcement actions if licensees do not meet the regulatory and safety standard discussed in the SRP. SRPs are also not inspection procedures. Inspectors can use SRPs as informational tools, but they should perform the inspection in accordance with the guidance contained in the applicable inspection procedure.

D.   Systematic Evaluation Program

In 1977, the NRC staff initiated the Systematic Evaluation Program (SEP) to review the designs of older operating nuclear power plants. These plants were licensed before the NRC issued its SRP in 1975. The SEP was divided into two phases. In Phase I, the staff defined 137 issues for which regulatory requirements had changed over time to warrant an evaluation of those plants licensed before the issuance of the SRP. In Phase II, the staff compared the design of a select number of the older plants to the SRP issued in 1975. Based on these reviews, the staff identified 27 of the original 137 issues that required some corrective action at one or more of the SEP plants that were reviewed. The staff referred to the issues on this smaller list as the SEP "lessons-learned" issues and concluded that the older operating plants that were not in the group of SEP plants examined (i.e., non-SEP plants) could improve their safety operations by taking corrective actions for these 27 issues. Therefore, the NRC staff concluded that inspectors should consider these 27 issues at the non-SEP plants to determine whether an adequate level of safety existed at these plants.

During inspections at the plants licensed before 1975, inspectors may find that the licensee's licensing basis is complex, and therefore, they may need to review numerous licensing documents to thoroughly evaluate the acceptability of particular issues of concern. Table 2 lists all current operating facilities reviewed as part of the SEP and those licensed before 1975.

Table 2. SEP and Pre-1975 Plants

SEP Plants	Pre-1975 Plants (non-SEP Plants)	
Palisades	Arkansas 1	Monticello
Ginna	Browns Ferry 1/2	Nine Mile Point 1
Oyster Creek	Brunswick 2	Oconee 1/2/3
Dresden 2	Calvert Cliffs 1	Peach Bottom 2/3
	Cook 1	Pilgrim
	Cooper	Point Beach 1/2
	Dresden 3	Prairie Island 1/2
	Duane Arnold	Quad Cities 1/2
	FitzPatrick	Robinson 2
	Fort Calhoun	Surry 1/2
	Hatch 1	Three Mile Island 1
	Indian Point 2/3	Turkey Point 3/4
	Kewaunee	Vermont Yankee
	Millstone 2	

Generic Letter (GL) 95-04, "Final Disposition of the Systematic Evaluation Program Lessons-Learned Issues," discusses the final disposition of the lessons-learned issues found in the SEP. References in GL 95-04 provide additional background information about the SEP.

E.   Safety Evaluation Reports

Safety evaluation reports (SERs) document the results of NRR reviews of proposed licensing actions, such as initial licensing reviews and subsequent license amendments. SERs are part of the licensing basis of the plant and are useful for inspectors to review so that the full scope of the plant's licensing basis is understood. SERs give insights into technical issues that a licensee has already resolved and can help inspectors focus the inspection.

F.   Technical Guidance—Manual Chapter Part 9900

Part 9900 of the NRC Inspection Manual contains technical guidance on various areas of interest to inspectors ranging from technical to licensing issues. This guidance was developed in specific areas to address NRC staff or industry questions in those areas.

As an example, technical guidance exists regarding Sections III and XI of the ASME BVP Code. Specifically, Code interpretations, use of engineering judgment, and flaw evaluations are discussed. Other notable areas where technical guidance is available include other Code-related issues, equipment repair issues, notices of enforcement discretion (NOEDs), TS guidance, treatment of degraded or nonconforming conditions, and equipment testing interpretation issues.

Typically, engineering-focused inspectors use the guidance on the treatment of degraded or nonconforming conditions along with the testing and Code guidance.

Field inspections and verifications are an important element of the NRC inspection program.

DESIGN AND LICENSING BASIS

A.   Design Basis (Bases)

Inspectors should recognize that the design basis is a subset of the licensing basis.

<u>RG 1.186, "Guidance and Examples for Identifying 10 CFR 50.2 Design Bases"</u>

RG 1.186 discusses the history of inadequate maintenance of design-basis documentation (DBD) by licensees and addresses the NRC's initiative requesting licensees to reconstitute their design-base information. The guide also endorses Appendix B, "Guidance and Examples for Identifying 10 CFR 50.2 Design Bases," to Nuclear Energy Institute (NEI) 97-04, "Design Bases Program Guidelines."

Design Basis
Design basis is the information that identifies the specific functions that a facility's SSC is to perform and the specific values or ranges of values chosen for controlling parameters as reference bounds for design. These values may be (1) restraints derived from generally accepted "state-of-the-art" practices for achieving functional goals or (2) requirements derived from analysis (based on calculations and/or experiments) of the effects of a postulated accident for which an SSC must meet its functional goals.

NEI 97-04 Appendix B, Guidance and Examples for Identifying 10 CFR 50.2 Design Bases

This document clarifies and expands the definition of design bases and discusses the relationship of the design bases to other 10 CFR requirements. It clarifies the definition of design bases to include those bounding conditions under which SSCs must perform design basis functions.

It further discusses how a facility's design bases can change. The design bases for a plant can change as a result of new NRC requirements after the approval of the original operating license and from changes to ensure compliance with NRC requirements. NEI 97-04 can be obtained from the NRC's Agencywide Documents Access and Management System (ADAMS Accession No. ML003771698).

Final and Updated Safety Analysis Reports

Final safety analysis reports (FSARs) and UFSARs can contain design- and licensing-basis information and describe the licensee's implementation of the general design criteria in Appendix A to 10 CFR Part 50. A license application originally includes the plant's <u>Preliminary Safety Analysis Report</u>. During the licensing stage, the licensee submits an FSAR. The licensee must describe the plant's individual safety systems in the FSAR in sufficient detail to allow the NRR reviewer to prepare an SER to support the granting of an operating license. As a result of routine updates after the granting of the license, the FSAR is often referred to as an UFSAR (or USAR). The UFSAR is both a licensing-basis document and a design-basis document because it describes how the plant was designed, constructed, maintained, and operated, but it does not by itself contain the regulatory requirements. The UFSAR is unique to each nuclear power plant station and, in the case of multiple-unit sites, to each unit of a station. UFSARs vary in the amount of detail they contain, which mostly depends on the age of the plant. Older plants may have as few as four volumes, whereas newer plants have many more.

Suggested Sources for Design-Bases Information

- UFSARs
- DBD
- system descriptions
- design calculations
- design analyses
- piping and instrumentation drawings
- significant design drawings
- significant surveillance procedures
- preoperational test documents
- vendor manuals

Locating Design-Basis Information

At the site, the UFSAR and the station DBD or system descriptions are good starting points for locating design-basis information. Because

Validation of engineering judgment and assumptions can provide insights into the adequacy of engineering products.

of the design-base reconstitution effort, most nuclear stations have a system DBD or System Description document for each safety-related system. Topical DBDs will address program areas such as fire protection and station blackout. These documents will address the design bases requirements for a system, describe components, often state actual limiting parameter values, and provide references that support design-basis information such as calculations, analyses, and testing documentation that support the design bases for that system and its components. The DBD usually lists licensing commitments related to the subject system or topical program.

B.   Licensing Basis

A facility's licensing basis is also referred to as the current licensing basis (CLB). The concept of the CLB was first introduced in the regulations through 10 CFR 50.54(f), as a result of agency considerations in the development of the NRC's backfit rule. The CLB again became an issue in the NRC's decisions on extending the licenses for plants beyond the original design life, and it is defined in 10 CFR 54.3, "Definitions." Although established in <u>10 CFR Part 54, "Requirements for Renewal of Operating Licenses for Nuclear Power Plants</u>," the definition represents the NRC's understanding of the scope of the CLB and generally applies to all reactor licensees.

10 CFR 54.3(a)

> The current licensing basis (CLB) is the set of NRC requirements applicable to a specific plant and a licensee's written commitments for ensuring compliance with and operation within applicable NRC requirements and the plant-specific design basis (including all modifications and additions to such commitments over the life of the license) that are docketed and in effect. The CLB includes the NRC regulations contained in 10 CFR Parts 2, 19, 20, 21, 26, 30, 40, 50, 51, 54, 55, 70, 72, 73, 100, and appendices thereto; orders; license conditions; exemptions; and technical specifications. It also includes the plant-specific design-basis information defined in 10 CFR 50.2 as documented in the most recent FSAR, as required by 10 CFR 50.71 and the licensee's commitments remaining in effect that were made in docketed licensing correspondence such as licensee responses to

NRC bulletins, generic letters, and enforcement actions, as well as licensee commitments documented in NRC safety evaluations or licensee event reports.

Facility Operating License

A facility's operating license describes the conditions and requirements for the operation of the nuclear unit, including the receipt, use, and possession of special nuclear material and byproducts. It also states the maximum authorized power level for power operation. The license includes appendices, such as the TS.

Appendix A to 10 CFR Part 50

Appendix A, "General Design Criteria for Nuclear Power Plants," to 10 CFR Part 50, establishes the regulatory requirements for the design of nuclear power plants. The criteria are stated generally and grouped in the following categories:

I.   Overall Requirements
II. Protection by Multiple Fission Product Barriers
III. Protection and Reactivity Control Systems
IV. Fluid Systems
V.   Reactor Containment
VI. Fuel and Radioactivity Control

Because of the age of some licensed facilities, inspectors should verify that Appendix A to 10 CFR Part 50 applies to the facility being inspected.

Locating Current Licensing-Basis Information

Although licensing-basis information can sometimes be found in ADAMS, the word searches needed to find specific documentation can be time consuming and challenging. However, the licensee does have a collection of station-specific licensing-basis information in its database and should be able to provide inspectors with this information more readily than would a word search of ADAMS. The station DBDs or system descriptions usually list licensing-basis commitment references applicable to the subject system or topical program. The UFSAR is readily available and is also a source of some, but not all, licensing commitments.

The following are suggested sources of licensing-basis documentation:

Safety is paramount when conducting field walkdowns near high energy or hazardous plant areas.

- NRC Title 10 Federal regulations
- plants' TS
- UFSARs
- NRC SERs

- licensees' responses to GLs
- licensees' responses to notices of violation
- licensee event reports
- technical requirements manuals
- TS bases
- organizational topical reports
- QA plans
- fire protection reports
- offsite dose calculations
- physical security plans
- radiological emergency plans
- Core Operating Limits Report
- Pressure and Temperature Limits Report

There are various regulatory processes that include options for licensees to make changes to their licensing basis. These include 10 CFR 50.59, "Changes, Tests, and Experiments," and operating license conditions such as fire protection, quality assurance, emergency preparedness, and physical security plans; regulatory exemption requests; and license amendments under 10 CFR 50.90, "Application for Amendment of License, Construction Permit, or Early Site Permit." In 10 CFR 50.59, the NRC establishes the conditions under which licensees may make changes to the facility or procedures and may conduct tests or experiments without prior NRC approval. The NRC must review and approve the proposed changes, tests, and experiments that satisfy the definitions and one or more of the criteria in the regulation before licensees can implement them.

DESIGN CONTROL GUIDANCE DOCUMENTS

In 1967, the NRC's predecessor, the Atomic Energy Commission, published for comment general design criteria for light-water-cooled nuclear power plants (i.e., draft Appendix A to 10 CFR Part 50). Appendix A to 10 CFR Part 50 requires licensees to establish a QA program, whereas Appendix B to 10 CFR Part 50 defines the requirements of that program. These appendices were approved as final rules in 1971 and 1970, respectively. Before their issuance, the NRC held meetings and exchanged correspondence with applicants individually to resolve issues. The staff's positions eventually became RGs, which provide guidance on acceptable methods that licensees can use to meet regulations, but they are not regulations themselves.

As part of the licensing process, licensees must submit a QA plan to the NRC. Compliance with the QA plan is required because it is an attachment to the facility's operating license. The QA plan is structured to follow the requirements of Appendix B to 10 CFR Part 50, and it provides a general description of the established measures that licensees need to comply with Appendix B requirements, including Criterion III, "Design Control." As part of the preparation for an engineering-focused inspection, an inspector should periodically review the licensee's QA plan to better understand the licensee's design control measures. The inspector can further review design control implementing procedures during the inspection because they describe in greater detail how the licensee implements design control measures.

The American National Standards Institute (ANSI) published ANSI N45.2, "Quality Assurance Requirements for Nuclear Power Plants," to describe the requirements of Appendix B in greater detail. Industry standards, such as ANSI N45.2 and daughter standards (e.g., ANSI N45.2.11, "Quality Assurance Requirements for the Design of Nuclear Power Plants"), serve as "how-to" documents for the nuclear industry to meet NRC requirements, including Criterion III. The NRC endorsed this standard in RG 1.28, "Quality Assurance Program Requirements (Design and Construction)." Inspectors should review the licensee's QA plan, which will describe the established design control process, including commitments to industry standards; the SRP; safety analysis reports; other industry codes; regulations; and RGs to obtain a thorough understanding of the licensee's design control process. ANSI (sometimes coded as ASME) standards may be obtained through the NRC's electronic Technical Library.

RG 1.28, "Quality Assurance Program Requirements (Design and Construction)"

In RG 1.28, Revision 3, the NRC endorsed industry standard ASME NQA-1-1983, "Quality Assurance Program Requirements for Nuclear Facilities." The standard mirrors, and in some cases provides greater guidance than that in ANSI N45.2.11 as an acceptable method that licensees can use to implement QA programs during the design and construction phases of a nuclear power plant. Inspectors should remain informed with respect to the RG position and the specific licensee commitments in this area. Additionally, the NRC may later endorse more recently issued versions of the ASME standard, as reflected in draft RGs.

ASME NQA-1-1983 has three main sections: (1) Basic Requirements, (2) Supplements, and (3) Appendices. The Basic Requirements section provides the basic requirements for establishing and executing QA programs. The Supplements section amplifies the individual requirements of the Basic Requirements section. The Appendices section provides non-mandatory guidance for meeting the Basic Requirements and Supplements sections.

ANSI N45.2.11, "Quality Assurance Requirements for the Design of Nuclear Power Plants"

ANSI N45.2.11 describes the minimum QA requirements that licensees must implement during the design of nuclear power plant SSCs. The SSCs are those that are required to prevent accidents that could cause undue risk to the health and safety of the public or those that are required to mitigate the consequences of an accident. During the licensing phase of nuclear plants, most licensees commit to following the guidance contained in the standard. As is the case with all licensing actions, inspectors should verify actual licensee commitments through a review of the facility's UFSAR and other licensing-basis documents. Licensees structure their QA program regarding design control to incorporate the guidance contained in the standard and to meet the requirements of Criterion III of Appendix B to 10 CFR Part 50. The standard itself covers various elements of an effective design control program.

A multi-discipline team inspection approach is warranted for complex plant modifications.

A brief discussion of salient portions of ANSI N45.2.11 follows. The discussion summarizes the guidance contained within the standard. Inspectors should periodically review the entire standard to develop a comprehensive understanding of all design control standard elements.

Design Process and Input Requirements

Criterion III of Appendix B to 10 CFR Part 50 requires, in part, the following:

> Measures shall be established to assure that applicable regulatory requirements and the design basis, as defined in 10 CFR 50.2 and as specified in the licensee application, for those structures, systems, and components to which this Appendix applies are correctly translated into specifications, drawings, procedures, and instructions.

Sections 3 and 4 of ANSI N45.2.11 provide the following guidance on this criterion:

> Applicable design inputs, such as design bases, regulatory requirements, codes and standards, shall be identified, documented, and their selection reviewed and approved.

> Design activities shall be prescribed and accomplished in accordance with procedures of a type sufficient to assure that applicable design inputs are correctly translated into specifications, drawings, procedures, or instructions.

Inspectors should recognize that design input requirements include the following (but the list is not all inclusive):

- basic functions of SSCs
- performance requirements such as capacity, rating, and system output
- codes and standards
- design conditions, such as pressure, temperature, and voltage
- loads, such as seismic, thermal, and dynamic
- environmental conditions anticipated during operation
- operational requirements under various plant conditions

Additionally, licensees are to develop procedures that include requirements for the control of design analyses (calculations). Licensees must identify calculations by subject, originator, reviewer, and date or other means to assure calculations are retrievable. Calculations should document known and unverified assumptions.

Design Verification

Criterion III of Appendix B to 10 CFR Part 50 requires, in part, the following:

> The design control measures shall provide for verifying or checking the adequacy of design, such as by the performance of design reviews, by the use of alternate or simplified calculational methods, or by the performance of a suitable testing program. The verifying or checking process shall be performed by individuals or groups other than those who performed the original design, but who may be from the same organization. Where a test program is used to verify the adequacy of a specific design feature in lieu of other verifying or checking processes, it shall include suitable qualification testing of a prototype unit under the most adverse design conditions.

Section 6 of ANSI N45.2.11 provides guidance on "verifying or checking the adequacy of design":

> Measures shall be established to verify the adequacy of design. Design verification is the process of reviewing, confirming, or substantiating the design but one or more methods to provide assurance that the design meets the specified design inputs.

Walkdowns in sensitive plant areas may require coordination with licensee staff in Operations or Plant Safety.

> Design verification shall be performed by any competent individuals or groups other than those who performed the original design but who may be from the same organization.

Cursory supervisory reviews do not satisfy the intent of this standard. Design verification may vary from spot checking of calculations to actual tests in the field.

The extent of the design verification required is a function of the importance to safety of the item under consideration, the complexity of the design, the degree of standardization, the state-of-the-art, and the similarity with previously proven designs.

Acceptable verification methods include, but are not limited to, design reviews, alternate calculations, qualification testing.

For design reviews, "the depth of review can range from a detailed check of the complete design to a limited check of such things as the design approach and the results obtained." For alternate calculations, verification can be accomplished by performing simplified hand calculations. For qualification testing, "all pertinent operating modes shall be considered in determining the design conditions where it is intended that the test program confirm the adequacy of the overall design."

Design Change Control

Criterion III of Appendix B to 10 CFR Part 50 requires, in part, the following:

Design changes, including field changes, shall be subject to design control measures commensurate with those applied to the original design and be approved by the organization that performed the original design unless the applicant designates another responsible organization.

Section 8 of ANSI N45.2.11 provides guidance on "design changes":

Documented procedures shall be provided for design changes to approved design documents, including field changes…. The changes shall be justified and subjected to design control measures commensurate with those applied to the original design.

CRITERION III OF APPENDIX B TO 10 CFR PART 50 CROSS-REFERENCE TABLE

Table 3 identifies specific RGs and ANSI standards that provide regulatory guidance with respect to specific requirements in Criterion III of Appendix B to 10 CFR Part 50. The format in Table 3 can serve as a quick reference for inspectors.

Table 3. Criterion III of Appendix B to 10 CFR Part 50 Cross-Reference Table

	Requirement	RG	ANSI Standard
III.1	Measures shall be established to assure that applicable regulatory requirements and the design basis, as defined in 10 CFR 50.2 and as specified in the licensee application, for those structures, systems, and components to which this Appendix applies are correctly translated into specifications, drawings, procedures, and instructions.	1.28	N45.2 N45.2.10 N45.2.11
III.2	These measures shall include revisions to assure that appropriate quality standards are specified and included in design documents and that deviations from such standards are controlled.	1.26 1.28 1.29 1.75	N45.2 N45.2.10 N45.2.11 IEEE 279
III.3	Measures shall also be established for the selection and review for suitability of application of materials, parts, equipment, and processes that are essential to the safety-related functions of the structures, systems, and components.	1.28	N45.2 N45.2.10 N45.2.11
III.4	Measures shall be established for the identification and control of design interfaces and for coordination among participating design organizations.	1.28	N45.2 N45.2.10 N45.2.11
III.5	These measures shall include the establishment of procedures among participating design organizations for the review, approval, release, distribution, and revision of documents involving design interfaces.	1.28	N45.2 N45.2.10 N45.2.11
III.6	The design control measures shall provide for verifying or checking the adequacy of design, such as by the performance of design reviews, by the use of alternate or simplified calculational methods, or by the performance of a suitable testing program.	1.28 1.89	N45.2 N45.2.10 N45.2.11 IEEE 323

III.7	The verifying or checking process shall be performed by individuals or groups other than those who performed the original design, but who may be from the same organization.	1.28	N45.2 N45.2.10 N45.2.11
III.8	Where a test program is used to verify the adequacy of a specific design feature in lieu of other verifying or checking processes, it shall include suitable qualification testing of a prototype unit under the most adverse design conditions.	1.28 1.40 1.73 1.89 1.100	N45.2 N45.2.10 N45.2.11 IEEE 334 IEEE 382
III.9	Design control measures shall be applied to items such as the following: reactor physics, stress, thermal hydraulic, and accident analysis; compatibility of materials; accessibility for inservice inspection, maintenance and repair; the delineation of acceptance criteria for inspections and tests.	1.28	N45.2 N45.2.10 N45.2.11
III.10	Design changes, including field changes, shall be subject to design control measures commensurate with those applied to the original design and be approved by the organization that performed the original design unless the applicant designates another responsible organization.	1.28	N45.2 N45.2.10 N45.2.11

PLANT ENGINEERING PRODUCTS

A. Overview

Engineers play a vital role in the design, maintenance, and operation of commercial nuclear power plants. Engineering work products include, but are not limited to, the following:

- calculations
- modifications
- temporary modifications
- operability determinations
- procedures
- corrective actions
- dedication of commercial-grade parts for safety-related applications

B. Calculations

A facility's operating license requires licensees to implement their operational QA program. As discussed earlier, ANSI N45.2.11 parallels the design control requirements and commitments in most QA program manuals. In general, the ANSI standard specifies that licensees take the following seven actions:

(1) Activities affecting design control shall be specified and accomplished in accordance with approved procedures.

(2) Responsibilities shall be specified in procedures.

(3) Applicable design inputs, such as design bases, regulatory requirements, and codes and standards, shall be identified.

(4) Design inputs shall include, where applicable, the following:

- the functions of each SSC
- performance requirements, such as capacity, rating, and system output
- industry codes, industry standards, and regulatory requirements
- design conditions, such as pressure, temperature, fluid chemistry, and voltage
- loads, such as seismic, wind, thermal, and dynamic
- environmental conditions anticipated during storage, construction, and operation and during a design-basis accident (e.g., pressure, temperature, humidity, corrosiveness, site elevation, wind direct, nuclear radiation, electromagnetic radiation, and duration of exposure)
- interface requirements
- material requirements
- mechanical requirements, such as vibration, stress, shock, and reaction forces
- structural requirements covering such items as equipment foundations and pipe supports

- hydraulic requirements, such as pump net positive suction head (NPSH), allowable pressure drop, and allowable fluid velocity
- chemistry requirements
- electrical requirements, such as source of power, voltage, raceway requirements, electrical insulation, and motor requirements
- layout and arrangement requirements
- operational requirements under various conditions, such as plant startup, normal plant operation, plant shutdown, plant emergency operations, special or infrequent operations, and system abnormal or emergency operations
- instrumentation and control requirements, including controls and alarms
- access and administrative control requirements for security
- redundancy, diversity, and separation requirements (single-failure proof)
- identification of those events and accidents against which the SSCs must be designed to withstand

(5) Design analysis shall be performed in a planned and controlled manner. Analyses shall be sufficiently detailed as to purpose, method, assumptions, design input, references, and units such that a person technically qualified in the subject can review and understand the design analysis and verify its adequacy.

(6) Assumptions and references shall be documented.

(7) Design verification may include design reviews, alternate calculations, or qualification testing, as follows:

- **Design Reviews:** Design reviews are critical reviews to ensure that design documents, such as drawings, calculations, analyses, or specifications, are correct and satisfactory. Design reviews can range from multiple-organization reviews to single-person reviews. The results of the reviews shall be documented.

- **Alternate Calculations:** The use of alternate methods of calculations or analyses can help verify, through comparison, some types of calculations or analyses. A person or persons other than those who performed the original calculations shall perform these alternate methods. The alternate method used may be a more simplified approach (e.g., using a hand calculation to check a computer code output). A simplified method must provide results that are consistent with the original calculations or analyses.

- **Qualification Testing:** Design verification for some design or specific design features can be achieved by suitable qualification testing of a prototype or initial production unit. The testing shall demonstrate the adequacy of the performance of a design under the most adverse design conditions.

The review and inspection of engineering calculations are important inspection activities. Calculations will exist in various forms and for different purposes. In general, most calculations are documented in a manner to ensure that applicable design control requirements are met. Specifically, 10 CFR 50, Appendix B, Criterion III requires, in part, measures for verifying or checking the adequacy of a design. A licensee may accomplish this through design reviews, alternate calculations, and/or qualification testing. If qualification testing is used to demonstrate the adequacy of a design, the testing of a prototype unit must be performed under the most adverse anticipated conditions.

In addition, calculations will often validate the quantitative values contained in the TS, USAR, and other commitments to the NRC. Of critical importance is the consistent relationship between design calculations, operational surveillances, procedures, and the licensing basis (i.e., NRC commitments, TS, and FSARs). Once minimum thresholds of performance are identified, procedures may use acceptance criteria that are more conservative, but not less conservative, than those specified in calculations.

A typical calculation may have the following sections:

• **Purpose:** This section discusses the reason for the calculation.

• **Assumptions:** This section discusses the starting points for the calculation. Inspectors should verify that assumptions are consistent with the licensing-basis and engineering-reference documents. Inspectors should discuss questionable assumptions with the licensee staff to fully evaluate their appropriateness, and, where possible, the assumptions should be independently verified. Fundamentally, the stated assumptions define the boundaries for the calculation and generally make the calculation "simpler." The inspector should focus on whether the assumption is valid and whether it could have a significant impact on the results of the calculation.

   Inspectors should verify that the licensee's "bounding analysis" approach truly bounds the worst-case scenarios and events. Often, licensees evaluate Loss of Coolant Accident scenarios as worst-case events, but occasionally "smaller" events provide the most limiting system conditions and demands.

• **References:** Calculation references are an important element of a good calculation. Recently published papers are not necessarily the most appropriate references—just as time-proven "industry classics" are not necessarily valid either. Inspectors should consider whether the NRC has endorsed the stated reference in an RG, NUREG, or other regulatory document. Absence of this endorsement does not necessarily make the reference inappropriate for use. However, the NRC has found that some standards endorsed by bona fide organizations (e.g., ASME) do not provide acceptable guidance. Agency endorsement of a code or standard provides a high level of confidence in the applicable guidance. Conversely, licensees are not expected to follow all standards. Inspectors have little regulatory leverage to require licensees to follow a code or standard where there is no corresponding licensee commitment or specific NRC regulation. In fact, imposing such a requirement upon licensees could constitute a regulatory backfit (see 10 CFR 50.109, "Backfitting," and associated agency guidance documents regarding the backfit process).

• **Design Inputs:** An important consideration in a calculation is the inputs. Inputs can include licensing-basis information (e.g., minimum flow rates, voltages, operating, and accident conditions) and system drawings. Inspectors should verify the accuracy of the calculation inputs, as practical as possible, and should develop a good understanding of the pertinent system drawings before and during the review of the calculation.

• **Analysis:** Engineering theory and the mathematics of the methodology are two distinct elements that inspectors should consider when reviewing the technical adequacy of an engineering calculation.

Inspectors should verify the validity of the engineering theory. For instance, many calculational formulas are valid only under certain conditions. The inspectors should verify that the applicable calculational limitations are properly considered. Additionally, inspectors should often perform independent, simplified calculations to verify that the licensee's calculational result is correct.

Criterion III, "Design Control," requires that "design control measures shall provide for verifying or checking the adequacy of design, such as by the performance of design reviews, by the use of alternate or simplified calculational methods, or by the performance of a suitable testing program...." Inspectors should be able to identify and verify the method by which the calculation was independently verified.

The mathematics of the methodology should also be independently checked, where possible. Plant engineers are not immune to making errors in their computations.

Additionally, the licensee's calculation should also have considered defense-in-depth elements. These include single failure, common-mode failure, separation (physical and electrical), and redundancy, where applicable. The facility's licensing-basis documents will discuss the applicability of these elements to particular systems.

- **Conclusions:** At this stage in the inspection process, the inspector should review the calculation conclusions and consider whether the calculation results seem reasonable or whether they conflict with other existing approved calculations. The inspector should focus on performing critical reviews of the calculations and on verifying their adequacy.

C. Modifications and Temporary Modifications

Plant modifications are one of the major engineering work products throughout the life of the plant. Modifications are implemented for various reasons, including to satisfy new regulatory requirements (e.g., station blackout and Appendix R), to improve the reliability of SSCs, to resolve degraded or nonconforming SSCs, and to improve plant efficiency.

Plant modifications are important because these engineering activities can affect the facility's conformance to the design and licensing basis. Control of design basis and plant configuration is important to ensure that the plant's design, operation, maintenance, and modifications remain consistent with the facility's FSAR and other licensing-basis documents.

The difference between a modification and a temporary modification is the duration of the change. Modifications are permanent changes, whereas temporary modifications are invoked

Observation of in-progress activities provides opportunities to objectively assess licensee performance.

for a limited duration. Design control requirements apply to all forms of modifications, including temporary modifications. The requirements in 10 CFR 50.59 also apply if an SSC is described in the FSAR.

Examples of modifications include, but are not limited to, the following:

- increasing the size of a motor operator
- changing the gear ratio on a motor operator
- installing a different pump into an existing system
- installing a new type of breaker into an existing breaker cubicle

Temporary modifications tend to be less invasive than permanent modifications, but they still can be significant. Typical temporary modifications include the following:

- installing temporary power sources to non-safety-related buses during an outage
- installing a temporary pressure gauge
- installing temporary temperature instruments
- lifting leads on nuisance alarms in the control room

For changes to the facility that do not last longer than the next outage, refer to Regulatory Issue Summary (RIS) 2005-20, "Revision to NRC Inspection Manual, Part 9900: Technical Guidance, 'Operability Determinations & Functionality Assessments for Resolution of Degraded or Nonconforming Conditions Adverse to Quality or Safety'," for important guidance with respect to operability and 10 CFR 50.59.

D. Operability Determinations

In many cases, a licensee will identify that an SSC is in a degraded condition or a nonconforming condition (i.e., not in compliance with an accepted code or standard). The inspectors should verify that the licensee has entered the concern into their corrective action program. In some cases, an engineering evaluation will be performed to support an operability determination. This assessment may be qualitative or quantitative (i.e., a calculation). The inspectors should evaluate the validity of any assumptions and the accuracy of calculations.

Agency expectations with respect to SSC operability are contained in RIS 2005-20 and its associated attachment, NRC Inspection Manual, Part 9900, "Technical Guidance, 'Operability Determinations & Functionality Assessments for Resolution of Degraded and Nonconforming Conditions Adverse to Quality or Safety." Inspectors should be familiar with this guidance. Ensuring that licensees perform adequate operability evaluations is an important responsibility of NRC inspectors. The following high-level concepts are provided:

Verification of engineering inputs and assumptions is critically important.

- **Technical Specification Compliance:** If a licensee is not meeting a TS requirement, the licensee must implement the applicable TS action requirements. In some cases, the licensee may be able to show that the risk to the public is not compromised by failure to follow the TS action statement. In those cases, the licensee may request a NOED to allow a temporary noncompliance with the TS action statement. These requests require NRC approval. Specific guidance regarding NOEDs is located in the NRC Inspection Manual, Part 9900, "Technical Guidance, 'Operations - Notices of Enforcement Discretion."

- **Operable and Operability:** A system, subsystem, train, component, or device shall be "operable" or have "operability" when it is capable of performing its specified function(s) and when all necessary attendant instrumentation, controls, electrical power, cooling or seal water, lubrication, and other auxiliary equipment that are required for the system, subsystem, train component, or device to perform its function(s) are also capable of performing their related support functions.

It is important to stress that the SSC must be capable of performing its safety function for the required licensing-basis mission time (which may be different than that of the probabilistic risk analysis).

- **Inoperable:** Inspectors should recognize that under certain circumstances, an SSC can be "inoperable" even if it meets its TS surveillance requirements. For example, a diesel generator could pass a TS surveillance load test (i.e., carrying the required kilowatts) while operating with high cylinder temperatures. In this case, the TS-required surveillance testing does not by itself envelop the degraded or nonconforming condition (i.e., high temperatures). Upon discovery of a degraded or nonconforming condition, the licensee must evaluate the condition and make an operability determination regarding whether the equipment can perform the intended design function under worst-case anticipated conditions for the duration of the mission time.

E. Procedures

Regulations, such as Criterion V, "Instructions, Procedures, and Drawings," of Appendix B to 10 CFR Part 50 and the administrative sections of the TS, require that all safety-related activities be performed in accordance with approved procedures and work instructions. Engineers write surveillance and test procedures to verify that SSCs continue to perform satisfactorily in service. Engineers also write post--modification test instructions to ensure that the modification did not adversely affect the performance of SSCs. When reviewing surveillance and test procedures, inspectors should, as a minimum, perform the following actions:

- Verify that test procedure acceptance criteria are consistent with the licensing basis (e.g., FSARs and TS).

- Verify that the tests are performed without undue preconditioning. Review guidance contained in NRC Inspection Manual, Part 9900, "Technical Guidance, Maintenance - Preconditioning of Structures, Systems, and Components Before Determining Operability." Some preconditioning is allowed in a few limited instances (e.g., emergency diesel generator testing).

- To the extent practicable, verify that testing reasonably verifies that SSCs will perform satisfactorily under design-basis (worst-case) accident conditions. For example, licensees will not likely be able to test heat exchangers under worst-case conditions; therefore, they must perform an extrapolation of the test data to verify design-basis capabilities.

- Verify that the instruments used in surveillances are calibrated (see Criterion XII, "Control of Measuring and Test Equipment," of Appendix B to 10 CFR Part 50). Test instrument uncertainties must be accounted for either explicitly (i.e., in a calculation) or implicitly (i.e., there is obviously sufficient margin so that a rigorous calculation is not needed).

F.   Corrective Action Determinations

Plant engineers may specify corrective measures to correct degraded or nonconforming conditions. In accordance with Criterion XVI, "Corrective Action," of Appendix B to 10 CFR Part 50, the corrective measures must be prompt and correct the condition adverse to quality. In addition, for significant conditions adverse to quality, the licensee must identify the cause of the condition and take sufficient corrective measures to prevent repetition. Equipment failures (e.g., emergency diesel generator and ECCS pump failures) are significant conditions adverse to quality.

G.   Dedication of Commercial-Grade Parts for Safety-Related Applications

In some instances, safety-grade replacement components are no longer available for purchase. As components wear out and need replacement, licensees will obtain commercially available replacement parts and will need to qualify the components for safety-related applications. When this is done, licensees must take the following actions:

- Assume responsibility for reporting pursuant to 10 CFR Part 21, "Reporting of Defects and Noncompliance."
- Ensure that QA program, seismic, and environmental qualifications requirements are met.
- Establish a means to verify that the component has the ability to perform design-basis conditions (e.g., identify safety functions and perform qualification testing, post-maintenance testing, and in-service testing).
- Establish necessary controls to ensure that the component will remain operable.
- Verify that the vendor has acceptable QA practices.
- Establish storage and preventive maintenance requirements.
- Establish a service life for the components.
- Perform periodic audits of the program.

ROBUST APPLICATION OF ENGINEERING PRINCIPLES

Inspectors should keep in mind that the overall intent of the design control process is to develop a well-supported solution to a defined problem affecting an SSC. The adequacy and completeness of the design documents are critically important elements and can be accomplished only through a well-defined and -controlled process. Effective implementation of such a process will provide greater assurance that safety- and risk-significant SSCs will be able to perform when required to ensure plant safety.

Often inspectors will review licensee conclusions and analyses and, after discussions with licensee engineers, will find that the engineering work products rely on "engineering judgment." Engineering judgment refers to technical judgments made by knowledgeable engineers experienced in the particular subject matter. The inspector should consider whether objective data exist to support the engineering judgment. Reliance on engineering judgment is predicated on the consistent and appropriate use of engineering "rules of thumb" and reliance on well-developed, understood, and widely accepted industry practices and standards. Engineering judgment does not mean that such conclusions cannot be questioned or are indisputable. When considering the appropriateness of engineering judgment, inspectors should consider the following questions:

> ### Design Margin
>
> How do you tell the difference between a risk analyst, a mathematician, a scientist, and an engineer?
>
> First, you ask them, "What does 2 plus 2 equal?"
>
> The risk analyst will respond, "Whatever number you would like it to be, but it has to be between 0 and 1."
>
> The mathematician will respond, "4."
>
> The scientist will respond, "The evidence points to 4."
>
> The engineer will respond, "Let's make it 5 to be on the safe side."

- Does the engineering judgment rely on assumptions or data that are not relevant to this issue (i.e., reliance on past successes to justify current assumptions)?
- Did the licensee consider other data that contradict the engineering judgment?
- Does the engineering judgment apply to "rules of thumb" alone?
- Does the conclusion seem reasonable?
- Does the conclusion account for industry operating experience?
- Is there conservatism incorporated in the engineer's conclusion?
- Is unsupported engineering judgment being relied on in lieu of testing to demonstrate design adequacy or system operability?

The licensee's engineering organization should recognize that engineers are guardians of design margin and represent the safety conscience of the plant. All engineering actions should reflect a thorough knowledge of systems and components, including applicable design and licensing basis. Engineering evaluations should be thorough and timely, and the conclusions reached should be based on sound engineering practices and reflect a robust application of engineering principles.

Engineering activities should reflect meticulous attention to detail to ensure that the plant's design basis and licensing commitments are maintained. Control of design basis and plant configuration is critically important to ensure that the plant's design, operation, maintenance, and modifications are consistent with the facility's FSAR and supporting documents.

As discussed in previous sections, rigorous design verifications and checking activities are a normal part of the design process. Design-basis documents should be continuously used to support engineering activities. The licensee's engineering organization should constantly monitor the plant's operation within the design bases.

Rigorous engineering activities should be characterized by an effort to ensure safe and correct actions rather than the removal of safety margins in an effort to justify the acceptability of existing degraded conditions.

VALUE-ADDED FINDINGS AND STARS DOCUMENTS

Regional staff often issue value-added findings (VAFs) and STARS documents as a method to communicate and share inspection findings and work techniques with other inspectors. These documents generally describe topics that provide additional benefits to other inspectors by describing a generic issue that may be applicable at other facilities or a unique inspection method or technique. Table 4 lists VAFs and STARS documents related to design control.

Table 4. VAFs and STARS Documents

Number	Title	Region/Site
2000-03	Potential Service Water Pump Inoperability Due to Forebay Level	Inadequate
2001-08	Modifications to Nonsafety-Related Systems Cause Safety-Related Inoperability	RIII-Prairie Island
2003-01	Inadequate Temporary Cooling for 480Vac Vital Switchgear	RI-Millstone
2004-17	Shutdown Service Water System Pipe Support Deficiencies	RIII-Clinton
2004-18	Inappropriate Use of a Common Nonsafety-Related Power Source To Feed Two Redundant Safeguard Electrical Control Circuits	RIII-Kewaunee
2005-14	Incorrect Sprinkler System for the Emergency Feedwater Pump Room	RIV-ANO
2005-15	Higher Room Temperatures Adversely Impacting Cable Ampacity Margins	RIII-Duane Arnold
2005-25	Failure To Perform Adequate Oversight of a Modification Performed by a Contractor	RIII-Dresden
2005-38	Strong Ownership of Issues Successfully Brings Closure to Old Issues	RIII-Duane Arnold
2005-41	Reactor Core Isolation Cooling Governor Not Qualified for Station Black Out Temperatures and Questionable Breaker Short Circuit Interrupting Capability	RIII-LaSalle
2005-42	Failure To Provide Electrical Coordination To Ensure That Fire-Induced Electrical Faults Would Not Result in the Loss of Post-Fire Alternative Safe Shutdown Equipment	RIII-Fermi
2006-03	Single Cell Charging Design Control	RIV-Columbia
2006-09	Containment Spray Suction Pipe Voiding	RIV-Wolf Creek
2006-10	Net Positive Suction Head of the Reactor Core Isolation Cooling Pump	RIII-LaSalle
2006-12	Inadequate Freeze Protection	RIV-Grand Gulf
2006-14	Inadequate Technical Review of Permanent Modification	RIII-Clinton
2006-18	Temporary Alteration Fails To Meet Design Analysis Assumptions	RI-Indian Point 2

REFERENCES

References discussed within this booklet are listed below. Users should thoroughly understand the licensee's licensing basis and commitments to understand the applicability of the reference documents. The industry standards discussed within the booklet are those which were generally in-use during the licensing phase of a significant number of nuclear power plants. Additionally, industry standards often are revised and the specific content may change; hence, specific revisions and dates of such standards, and other references, are omitted.

Code of Federal Regulations

10 CFR 21	"Reporting of Defects and Noncompliance"
10 CFR 50	"Domestic Licensing of Production and Utilization Facilities"
10 CFR 50.2	"Definitions"
10 CFR 50.34	"Contents of Applications; Technical Information"
10 CFR 50.54	"Conditions of Licenses"
10 CFR 50.55a	"Codes and Standards"
10 CFR 50.59	"Changes, Tests, and Experiments"
10 CFR 50.67	"Accident Source Term"
10 CFR 50.71	"Maintenance of Records, Making of Reports"
10 CFR 50.90	"Application for Amendment of License, Construction Permit, or Early Site Permit"
10 CFR 50.109	"Backfitting"
10 CFR 50 Appendix A	"General Design Criteria for Nuclear Power Plants"
10 CFR 50 Appendix B	"Quality Assurance Criteria for Nuclear Power Plants and Fuel Reprocessing Plants"
10 CFR 54	"Requirements for Renewal of Operating Licenses for Nuclear Power Plants"
10 CFR 54.3	"Definitions"
10 CFR 100.11	"Determination of Exclusion Area, Low Population Zone, and Population Center Distance"

NUREGs

NUREG-0800	"Standard Review Plan for the Review of Safety Analysis Reports for Nuclear Power Plants"
NUREG-1055	"Improving Quality and the Assurance of Quality in the Design and Construction of Nuclear Power Plants"
NUREG-1397	"An Assessment of Design Control Practices and Design Reconstitution Programs in the Nuclear Industry," February 1991

Regulatory Guides

RG 1.1   "Net Positive Suction Head for Emergency Core Cooling and Containment Heat Removal System Pumps" (Safety Guide 1)

RG 1.6   "Independence Between Redundant Standby (Onsite) Power Sources and Between Their Distribution Systems" (Safety Guide 6)

RG 1.9   "Application and Testing of Safety-Related Diesel Generators in Nuclear Power Plants"

RG 1.26   "Quality Group Classifications and Standards for Water-, Steam-, and Radioactive-Waste-Containing Components of Nuclear Power Plants"

RG 1.28   "Quality Assurance Program Requirements (Design and Construction)"

RG 1.29   "Seismic Design Classification"

RG 1.40   "Qualification Tests of Continuous-Duty Motors Installed Inside the Containment of Water-Cooled Nuclear Power Plants"

RG 1.47   "Bypassed and Inoperable Status Indication for Nuclear Power Plant Safety Systems"

RG 1.53   "Application of the Single-Failure Criterion to Safety Systems"

RG 1.63   "Electric Penetration Assemblies in Containment Structures for Nuclear Power Plants"

RG 1.73   "Qualification Tests of Electric Valve Operators Installed Inside the Containment of Nuclear Power Plants"

RG 1.75   "Criteria for Independence of Electrical Safety Systems"

RG 1.89   "Environmental Qualification of Certain Electric Equipment Important to Safety for Nuclear Power Plants"

RG 1.97   "Criteria for Accident Monitoring Instrumentation for Nuclear Power Plants"

RG 1.100   "Seismic Qualification of Electric and Mechanical Equipment for Nuclear Power Plants"

RG 1.106   "Thermal Overload Protection for Electric Motors on Motor-Operated Valves"

RG 1.147   "Inservice Inspection Code Case Acceptability, ASME Section XI, Division 1"

RG 1.186   "Guidance and Examples for Identifying 10 CFR 50.2 Design Bases"

RG 1.187   "Guidance for Implementation of 10 CFR 50.59, Changes, Tests, and Experiments"

RG 1.193   "ASME Code Cases Not Approved for Use"

Regulatory Issue Summary

RIS 2005-20   "Revision to NRC Inspection Manual, Part 9900 Technical Guidance, 'Operability Determinations & Functionality Assessments for Resolution of Degraded or Nonconforming Conditions Adverse to Quality or Safety'"

Information Notices

N 84-54   "Deficiencies in Design Base Documentation and Calculations Supporting Nuclear Power Plant Design"

IN 91-29   "Deficiencies Identified During Electrical Distribution System Functional Inspections"

IN 97-81   "Deficiencies in Failure Modes and Effects Analyses for Instrumentation and Control Systems"

IN 98-22   "Deficiencies Identified During NRC Design Inspections"

Generic Letters

GL 88-15   "Electric Power Systems—Inadequate Control Over Design Processes"

GL 91-05   "Licensee Commercial-Grade Procurement and Dedication Programs"

GL 92-03   "Compilation of the Current Licensing Basis"

GL 95-04   "Final Disposition of the Systematic Evaluation Program Lessons-Learned Issues"

Industry Standards and Guidance Documents

ANSI/ISA-67.04.01 "Setpoints for Nuclear Safety-Related Instrumentation"

ANSI/ASME N45.2 "Quality Assurance Program Requirements for Nuclear Facilities"

ANSI/ASME N45.2.10 "Quality Assurance Terms and Definitions"

ANSI/ASME N45.2.11"Quality Assurance Requirements for the Design of Nuclear Power Plants"

ANSI/ASME NQA-1 "Quality Assurance Program Requirements for Nuclear Facilities"

NEI 97-04   "Design Bases Program Guidelines"

NEI 96-07   "Guidelines for 50.59 Evaluations"

IEEE 279   "Criteria for Protection Systems for Nuclear Power Generating Stations"

IEEE 308   "Criteria for Class 1E Power Systems for Nuclear Power Generating Stations"

IEEE 323   Qualifying Class 1E Equipment for Nuclear Power Generating Stations"

IEEE 334   "Standard for Type Tests of Continuous-Duty Class 1E Motors for Nuclear Power Generating Stations"

IEEE 379   "Standard Application of the Single-Failure Criterion to Nuclear Power Generating Station Safety Systems"

IEEE 382   "Trial Use Guide for Type Test of Class 1E Electric Valve Operators for Nuclear Power Generating Stations"

IEEE 450   "Recommended Practice for Large Lead Storage Batteries for Generating Stations and Substations"

IEEE 484   "Recommended Practice for Installation Design and Installation of Large Lead Storage Batteries for Generating Stations and Substations"

IEEE 603   "IEEE Standard Criteria for Safety Systems for Nuclear Power Generating Stations"

ACRONYMS

ANSI	American National Standards Institute
ASME	American Society of Mechanical Engineers
BPV	Boiler and Pressure Vessel
CFR	Code of Federal Regulations
CLB	Current Licensing Basis
DBD	Design Basis Documentation
ECCS	Emergency Core Cooling Systems
FSAR	Final Safety Analysis Report
GL	Generic Letter
IEEE	Institute of Electrical and Electronics Engineers
NOED	Notice of Enforcement Discretion
NRC	U.S. Nuclear Regulatory Commission
NRR	Office of Nuclear Reactor Regulation
NRO	Office of New Reactors
NEI	Nuclear Energy Institute
NPSH	Net Positive Suction Head
NUREG	Nuclear Regulation
PSAR	Preliminary Safety Analysis Report
QA	Quality Assurance
RG	Regulatory Guide
RIS	Regulatory Issue Summary
ROP	Reactor Oversight Process
SEP	Systematic Evaluation Program
SER	Safety Evaluation Reports
SRP	Standard Review Plans
SSC	Structures, Systems, and Components
TS	Technical Specifications
UFSAR	Updated Final Safety Analysis Report
USAR	Updated Safety Analysis Report
VAF	Value Added Findings

NRC FORM 335 (9-2004) NRCMD 3.7	U.S. NUCLEAR REGULATORY COMMISSION	1. REPORT NUMBER (Assigned by NRC, Add Vol., Supp., Rev., and Addendum Numbers, if any.)
	BIBLIOGRAPHIC DATA SHEET *(See instructions on the reverse)*	NUREG-1913

2. TITLE AND SUBTITLE

DESIGN CONTROL

A Quick Reference Guide For NRC Inspectors

3. DATE REPORT PUBLISHED

MONTH	YEAR
August	2009

4. FIN OR GRANT NUMBER

5. AUTHOR(S)

Julio F. Lara, P.E.
Jennifer Tifft
Frank Arner
Randy Moore
George Replogle

6. TYPE OF REPORT

NUREG

7. PERIOD COVERED *(Inclusive Dates)*

8. PERFORMING ORGANIZATION - NAME AND ADDRESS *(If NRC, provide Division, Office or Region, U.S. Nuclear Regulatory Commission, and mailing address; if contractor, provide name and mailing address.)*

Region III
U. S. Nuclear Regulatory Commission
2443 Warrenvile Rd.
Lisle, IL 60532

9. SPONSORING ORGANIZATION - NAME AND ADDRESS *(If NRC, type "Same as above"; if contractor, provide NRC Division, Office or Region, U.S. Nuclear Regulatory Commission, and mailing address.)*

Same as above

10. SUPPLEMENTARY NOTES

11. ABSTRACT *(200 words or less)*

Knowledge management and transfer have become increasingly important as the U.S. Nuclear Regulatory Commission seeks to share the vast inspection knowledge of its experienced inspectors with those who have recently joined the agency. This booklet should serve as an inspection reference to further increase the understanding by, and development of, NRC inspectors who perform inspections in the areas covered by this booklet.

12. KEY WORDS/DESCRIPTORS *(List words or phrases that will assist researchers in locating the report.)*

Knowledge Management
Design Control
Quick Reference Guide

13. AVAILABILITY STATEMENT

unlimited

14. SECURITY CLASSIFICATION

(This Page)

unclassified

(This Report)

unclassified

15. NUMBER OF PAGES

16. PRICE

NOTES

NOTES

NOTES

AVAILABILITY OF REFERENCE MATERIALS
IN NRC PUBLICATIONS

NUREG-1913
August 2009

www.ingramcontent.com/pod-product-compliance
Lightning Source LLC
Chambersburg PA
CBHW081905170526
45167CB00007B/3154